Henry Carrington Bolton

The Smithsonian Institution

Its Origin, Growth and Activities

Henry Carrington Bolton

The Smithsonian Institution
Its Origin, Growth and Activities

ISBN/EAN: 9783337817091

Printed in Europe, USA, Canada, Australia, Japan

Cover: Foto ©berggeist007 / pixelio.de

More available books at **www.hansebooks.com**

THE SMITHSONIAN INSTITUTION.

ITS ORIGIN, GROWTH, AND ACTIVITIES.

BY

PROF. HENRY CARRINGTON BOLTON, PH. D.

REPRINTED FROM APPLETONS' POPULAR SCIENCE MONTHLY FOR JANUARY, 1896.

*Reprinted from Appletons' Popular Science Monthly
for January and February, 1896.*

THE SMITHSONIAN INSTITUTION.

ITS ORIGIN, GROWTH, AND ACTIVITIES.

By Prof. HENRY CARRINGTON BOLTON, Ph. D.

PART I.—ORIGIN OF THE INSTITUTION.

WHEN the packet Mediator, commanded by Captain Christopher H. Champlin, sailed into New York harbor on the 28th day of August, 1838, after a stormy voyage of forty-three days from London, it brought in its hold a legacy from an Englishman to the United States of America, which was intended and destined to benefit all mankind. This precious freight consisted of eleven boxes, containing one hundred and five bags, each bag containing one thousand gold sovereigns. The boxes were carefully landed and stored for safe keeping in the Bank of America; a few days later the gold was sent to the United States Mint at Philadelphia, where it was immediately recoined into American money, yielding $508,318.46. This magnificent sum was the bequest of James Smithson, Esq., F. R. S., to the United States of America.

We propose in these articles to consider the purpose of this bequest, the manner in which the United States administers it, and the benefits to mankind accruing therefrom.

JAMES SMITHSON was born in France, in the year 1765, of distinguished English parentage; as he himself wrote: "The best blood of England flows in my veins; on my father's side I am a Northumberlander, on my mother's I am related to kings."

Of Smithson's early life little is known. At Pembroke College, Oxford, the young man was an earnest student and showed a liking for scientific pursuits; he was especially proficient in chemistry, and spent his vacations in collecting ores and minerals for analysis. He was graduated on the 26th of May, 1786; and the impulse for scientific research gained at the university influenced

all his succeeding years. The highest ambition of an English man
of science is to append to his name the honorable initials F. R. S.,
and to enjoy the privileges accorded to Fellows of the Royal
Society. Recommended by Richard Kirwan, the Irish chemist,
Charles Blagden, the Secretary of the Society, Henry Cavendish,

JAMES SMITHSON AS AN OXFORD STUDENT, 1786.

the wealthy and eccentric physicist, and others, Smithson was
elected a Fellow exactly eleven months after leaving the university.

During his residence in London he cultivated the society of
authors, artists, and men of science. " His mind was filled with a
craving for intellectual development, and for the advancement of
human knowledge. To enlarge the domain of thought, to dis-
cover new truths, and to make practical application of these for
the promotion of civilization, were the great ends he had con-
stantly in view." Smithson possessed large means; he never married, and for family reasons preferred to live on the Continent,

spending most of his time in France, Italy, and Germany; in his constant journeys he made observations on the climate, physical features, geology, and industries of the regions visited. He formed collections of minerals, and, for convenience of analyzing them, traveled with a portable chemical laboratory.

Living on the Continent, he acquired a cosmopolitan character, and formed acquaintance with the leading *savants* of the time; among his friends and corrrespondents were Gay-Lussac, the chemist; Haüy, the mineralogist; Arago, the astronomer; Biot, the physicist, of France; Berzelius, the chemist, of Sweden; and Davy, Black, Wollaston, Cavendish, Thomson, Smithson Tennant, chemical philosophers, of England. If it is "by a man's position among his contemporaries and competitors that his work may most justly be appraised," Smithson's scientific attainments must be rated very highly.

Between the years 1791 and 1825 Smithson published twenty-seven scientific papers, of which eight appeared in the Philosophical Transactions of the Royal Society and nineteen in Thomson's Annals of Philosophy. These memoirs embrace a wide range of research: the first deals with the curious deposit in bamboo called *tabasheer*, which he proved to be "siliceous earth;" the second was a "Chemical Analysis of Some Calamines," in which he established a new mineral species, afterward named *smithsonite* by Beudant (1832). The larger number of his papers deal with chemistry applied to mineral analysis, but he also discussed the nature of vegetables and insects, the origin of the earth, the crystalline form of ice, and an improved method of making coffee. An examination of these contributions to knowledge shows that he was no mere dilettante in science, and that he carried on his researches in a philosophic spirit for the sake of truth; all his writings exhibit keen perception, concise language, and accurate expression.

Of Smithson's personal traits and social character very little is known; his dislike of publicity, his natural reserve, as well as his residence in foreign countries, separated him from friends who might have given us particulars. It is said that he frequently narrated an anecdote of himself which illustrated his remarkable skill in analyzing minute quantities of substances, an ability which rivaled that of Dr. Wollaston. Happening to observe a tear gliding down a lady's cheek, he endeavored to catch it on a crystal vessel; one half the tear-drop escaped, but he subjected the other half to reagents, and detected what was then called microcosmic salt, muriate of soda, and some other saline constituents held in solution.

James Smithson died at the age of sixty-four years, on the 27th of June, 1829, at Genoa, Italy, and was buried in the Protestant

cemetery near that city. His death occurred in the same year with
that of Davy, Wollaston, and Young, a fact mentioned by the
President of the Royal Society in announcing the loss of members.

About three years before his death, Smithson made a holo-
graphic will containing provisions of immense importance to

JAMES SMITHSON. (From a painting by Johnes, 1816.)

American science. After providing for an annuity to one faithful
old servant, and a benefaction to another, his will directed that
the whole of the income arising from his property of every kind
should be paid by his executors to his nephew, Henry James
Hungerford; and should his nephew have children the whole of
his property was bequeathed to them or their heirs after the
death of their father. In case, however, the nephew should die
without issue, Smithson provided as follows:

"I bequeath the whole of my property to the United States of
America, to found at Washington, under the name of the Smith-

sonian Institution, an establishment for the increase and diffusion of knowledge among men."

The motives which prompted James Smithson to bequeath his fortune to the young republic across the seas are not certainly known. In the year 1818 (or 1819) he had some misunderstanding with the Royal Society, owing to their refusal to print one of his papers, and from that date he published exclusively in Thomson's Annals of Philosophy; it is said that prior to this difficulty he had intended to make the Royal Society his legatee. Having, however, abandoned that plan, he seems to have perceived with a prophetic eye the "germs of rising grandeur" in the free American nation, and to have felt a desire to promote the increase and diffusion of knowledge in the New World.

Whether he was more friendly to republicanism than to monarchy, as some have claimed, is not certain; at all events, by selecting the United States of America as the depository of his trust "he paid the highest compliment to its intelligence and integrity, and testified his confidence in republican institutions and his faith in their perpetuity."

In attempting to fathom the thoughts which directed Smithson's attention to the United States we are met by the surprising fact that he had not a single correspondent or scientific friend in America, nor did he write a line in any of his papers indicating appreciation of the republic.

Mr. Hungerford survived his uncle only six years, during which he received the benefits of the will; he led an aimless, roving life on the Continent, and died at Pisa, Italy, June 5, 1835, under the name of Eunice de la Batut, this being the surname of his stepfather, a Frenchman whom Hungerford's mother had married. By this death the United States became entitled to the estate. The first intimation received by the Government to this effect arrived in a communication dated July 28, 1835, from the *chargé d'affaires* of the United States at London to the Secretary of State, transmitting a letter from the firm of attorneys who represented the bankers holding the estate in trust. The estate was estimated at £100,000. In December, President Andrew Jackson sent to Congress a message setting forth the facts in the case and asking for authority to accept the trust; in July, Congress passed an act authorizing the President to appoint an agent to prosecute in the Court of Chancery the right of the United States to the legacy. This simple measure was not, however, secured without great difficulty, being opposed by several active Congressmen. Mr. W. C. Preston, of South Carolina, thought the donation had been made partly with a view to immortalize the donor, and it was "too cheap a way of conferring immortality"; and Mr. John C. Calhoun, of the same State, was of the opinion that it was beneath the dignity of the United States to receive presents of this kind from any one. The bill was, however, supported by the Committee of the Judiciary, to which the matter had been referred, and advocated by Mr. James Buchanan, of Pennsylvania, Mr. Robert J. Walker, of Mississippi, and Mr. John Davis, of Massachusetts.

Under this act President Jackson appointed the Hon. Richard Rush, of Pennsylvania, agent to prosecute the claims of the United States. The selection of Mr. Rush was a very happy one: he had been Comptroller of the Treasury, Attorney-General, minister to England, and minister to France. He displayed integrity and ability, and a persistence which accomplished the end in view with unexampled dispatch. Beyond the usual delays incident to court procedure, Mr. Rush met with no difficulties save one. Madame Théodore de la Batut, the mother of Mr. Hungerford, presented a claim for a life interest in the estate of Smithson; and to expedite matters Mr. Rush agreed to a compromise, granting an annuity, which she enjoyed until her death in 1861.* As soon as the securities were transferred to Mr. Rush, he converted them into gold and shipped it to New York on the Mediator; accompanying the treasure were three boxes containing the personal effects of the testator, including his collection of minerals, library, etc. The money arising from the Smithson bequest was at first invested in State stocks, and on December 10, 1838, President Martin Van Buren announced to Congress the receipt and disposition of the legacy of James Smithson. In 1841 Arkansas having failed to pay interest, through the efforts of Hon. J. Q. Adams the funds were transferred to the Treasury of the United States, to bear interest at six per cent per annum.

Three years had been consumed in securing the legacy, and seven and a half years more were destined to pass before Congress

JOSEPH HENRY.

---

* The principal retained in England to meet this annuity was paid over to the Smithsonian Institution in 1864. This residuary legacy amounted to $26,210 (gold).

SMITHSONIAN INSTITUTION, WASHINGTON, D. C.

carried out the wishes of the testator by creating the Smithsonian Institution. To analyze the legislation during this period, to describe the many extraordinary schemes proposed, to merely name the Congressmen who were active in the prolonged discussion, would occupy more space than can be given to this entire article. Presidents Van Buren, Harrison, Tyler, and Polk came and went, each urging Congress to action, but the legislators suffered from the " embarrassment of riches " in a new sense. Among the plans prominently brought forward and considered at length were the following: Senator John Quincy Adams advocated an astronomical observatory; Senator Asher Robbins, of Rhode Island, favored the establishment of a National University; Senator Benjamin Tappan, of Ohio, proposed a botanical garden and an agricultural farm; Senator Rufus Choate, of Massachusetts, urged a grand library; Robert Dale Owen, of Indiana, preferred a normal school with lectureships on scientific subjects; Mr. Isaac H. Morse, of Louisiana, wanted the prizes awarded for the best written essay on ten subjects; and some legislators, wise in their own conceit, opposed every plan suggested. Mr. George W. Jones, of Tennessee, proposed that the whole fund be returned to any heirs at law or next of kin of James Smithson; and a similar disposition of the fund was advocated by Andrew Johnson, of Tennessee, and Mr. A. D. Sims, of South Carolina. It is interesting, in the light of later national events, to note the names of some of those who took part in these discussions: we find side by side the names of Jefferson Davis and Hannibal Hamlin, Andrew Johnson and Alexander H. Stephens, Howell Cobb and Stephen A. Douglas.

Meanwhile memorials from persons and institutions outside of Congress poured in, urging expedition, advocating particular bills and suggesting new plans. At least two societies of citizens sought to gain control of the magnificent fund which Congress was so slow in appropriating; the Agricultural Society of the United States, formed in the District of Columbia, memorialized Congress to apply the Smithsonian fund to its objects; and the National Institution for the Promotion of Science, organized in 1840 by representative men in Washington, sought union with or control of the embryonic establishment bearing Smithson's name. Dr. G. Brown Goode, in his Genesis of the United States National Museum (Report of the United States National Museum, 1891), points out that the President of this National Institution, Joel R. Poinsett, of South Carolina (Secretary of the Navy in 1840), deserves credit for introducing the feature of a national museum into the scheme for the Smithsonian Institution. Indeed, the organization of the Smithsonian Institution finally adopted bears marked resemblance to that of the National Institution both as regards the cast of officers and the objects of the establishmer'

But all attempts to merge the interests of the two bodies failed, partly owing to objections to placing the management of the new institution in the hands of a private corporation; meanwhile the National Institution changed its name to National Institute, but after a flourishing existence of five years it lost its power.

Although much deprecated at the time, the slowness with which Congress acted in disposing of Smithson's legacy had its advantages: weak schemes were exposed, public opinion was educated, and the judgment of Congress itself was elevated by the prolonged discussions. The broad provisions of the will, open to the charge of vagueness, gave scope to the variety of views we have named and furnished ground for the delay. It is interesting to note that the act creating the Smithsonian Institution, adopted August 10, 1846, embodies nearly all the best features of the numerous schemes proposed during the ten years which had elapsed.

The act of incorporation was the work of many minds and to some extent a compromise; no one person should receive credit for its provisions, but mention should be made of Senator Benjamin Tappan, Robert Dale Owen, and William J. Hough, who drew up the bill eventually agreed upon. Stripped of legal verbiage and condensed, the bill is as follows:

TITLE.—A bill to establish the "Smithsonian Institution" for the increase and diffusion of knowledge among men.

Preamble : Rehearses the facts as to Smithson's bequest and the acceptance by the United States, and directs that the President and Vice-President of the United States, the Secretary of State, the Secretary of the Treasury, the Secretary of War, the Secretary of the Navy, the Postmaster-General, the Attorney-General, the Chief Justice, and the Commissioner of the Patent Office of the United States, and the Mayor of the city of Washington, during the time for which they shall hold their respective offices, and such other persons as they may elect honorary members, be constituted an "establishment" by the name of the Smithsonian Institution.

Section 2 provides for investment of the Smithson fund and payment of the interest thereon; also appropriates a sum for erection of a suitable building.

Section 3 provides that the business of said institution shall be conducted at the city of Washington by a Board of Regents to be composed of the Vice-President of the United States, the Chief Justice, and the Mayor of the city of Washington, together with three members of the Senate and three members of the House of Representatives, and six other persons, two of whom shall be members of the National Institute. The act then provides for the manner of appointment, the time of service, the filling of vacancies, the election of a Chancellor and Secretary by the Board of

Regents and of an executive committee, as well as for the payment of money needed for conducting the institution; also, an annual report to be submitted to Congress.

Section 4 provides for the selection of a suitable site for a building.

Section 5 provides for the erection of a building of plain and durable materials, of sufficient size for rooms to contain objects of natural history, including a geological and mineralogical cabinet, a chemical laboratory, a library, a gallery of art, and the necessary lecture rooms; also provides for the expense of this building.

Section 6 enacts that in proportion as suitable arrangements can be made for their reception all objects of art and of foreign and curious research, and all objects of natural history, plants, geological and mineralogical specimens, belonging or hereafter to belong to the United States which may be in the city of Washington shall be arranged as best to facilitate their examination and study in the building to be erected; also new specimens to be so arranged; also minerals, books, and other property of James Smithson to be preserved in the institution.

Section 7 enacts that the Secretary of the Board of Regents shall take charge of the building and contents, shall discharge the duties of librarian and of keeper of the museum, and may employ assistants, and provides for their compensation.

Section 8 provides for meetings at which the President or Vice-President of the United States shall preside, and appropriates a sum not exceeding twenty-five thousand dollars annually for the formation of a library.

Section 9 enacts that moneys accrued as interest upon the fund, not herein appropriated, may be disposed of by the Board of Regents as they direct.

Section 10 enacts that one copy of all copyrighted books, engravings, maps, etc., shall be sent to the Librarian of the Smithsonian Institution, and one to the Librarian of Congress.

Section 11 gives to Congress the right to amend any of the provisions of this act.

This act was signed by President James K. Polk, August 10, 1846. It embodies the features of a national museum, a library, with provisions for copyrighted books, an art gallery, and lecture rooms, presumably for scientific courses though no special provision for them is made. It places the executive work in the hands of a Secretary, and the general oversight with care of finances in the power of a Board of Regents, which board includes the highest officials in the Government of the United States.

The opponents of this bill, though defeated, still endeavored to change its character. Eighteen months after its passage, Andrew Johnson, of Tennessee, introduced a bill to change the Smithsonian Institution to "Washington University, for the Benefit of Indigent Children of the District of Columbia," and spoke in favor of remodeling the entire plan so as to convert the institution into a university to include the manual-labor feature, mechanic arts, and agriculture. Mr. Embree wanted at the same time to graft upon the institution a department for collecting and arranging information on agriculture, common-school education, political economy, and the useful arts and sciences, which information shall be published and circulated gratuitously among the people.

These attempts to tinker with the act of incorporation received their quietus on August 8, 1848, when the House of Representatives adopted a resolution to the effect that it is inexpedient to change and modify the act in the manner proposed. In 1878, and again in 1894, the act of incorporation was revised and somewhat simplified; the two Regents were no longer to be chosen from members of the National Institute, which meanwhile had died, and other slight changes were made.

SPENCER F. BAIRD.

Congress having appointed Regents, they organized by electing a Chancellor and temporary secretary. The act of incorporation placed great responsibilities in the secretary's office, and the Regents felt that the advancement of the proper interests of the trust made it essential that the Secretary of the Smithsonian Institution should be a man possessing weight of character and a high grade of talent; that he also possess eminent scientific and general acquirements; that he be capable of advancing science and promoting letters by original search and effort, and well qualified to act as a respected channel of communication between the institution and scientific and literary individuals in this and foreign countries. To this important position the Regents invited Prof. Joseph Henry, of the College of New Jersey, widely known in both hemispheres by his splendid discoveries in electro-magnetism and universally respected as a man by all who knew him. His acceptance of the secretaryship was a most fortunate event for the institution, insuring its high scientific standard, its wise and economical administration, and its superior reputation at home and abroad. Henry's Programme of Organization, presented to the Board of Regents December 8, 1847, is a model of skillful analytical statement, proposing plans for the increase of knowledge and its diffusion among men; in it he laid down broad lines of action and established the foundations on which the existing edifice stands. Henry devoted the rest of his life, thirty-three years, to the development of this programme, and the institution owes to him an everlasting debt of gratitude for his enlightened, pure, and able administration of the trust.

After the plans of Mr. James Renwick, Jr., for a Norman building, had been accepted, its erection in the Mall was conducted slowly, being completed in 1855, at an expense of about three hundred and fourteen thousand dollars. Meanwhile prudent economy in expenditures enabled Henry to add one hundred and fifty thousand dollars of accrued interest to the original fund.

A library was begun by exchange and purchase, and materials for a museum collected and housed. Besides these interests, the institution adopted the plan of promoting original research by assisting men of science in their labors; at the same time series of investigation were instituted, explorations conducted, and the results of all these endeavors were published and distributed to all the learned societies and important libraries throughout the world.

Whenever a man was found capable of adding to the sum of human knowledge, the institution assisted him by supplying books not otherwise attainable, instru-

S. P. LANGLEY.

ments of research, specimens of materials, and objects under investigation, and in some instances special grants of money were made for personal expenses. The specimens in all branches of natural history were not confined to the glass cases of the museum, but freely loaned to men engaged in special lines of research; and if the specimens required were not on hand, the institution undertook to obtain and to supply them, the only return asked for being

that full credit be given to the name of Smithson. This liberal policy has never been discontinued.

The institution established systematic meteorological observations, it instituted the first telegraphic weather service, published meteorological tables and charts, and became, in fact, the parent of the present Weather Bureau.

The institution early adopted a policy of doing nothing which could be accomplished as well by other means, and of relinquishing undertakings causing a draft upon its finances so soon as other bodies, or the Government, should agree to take them in charge. In pursuance of this wise plan the Secretary and the Regents induced Congress from time to time to make separate appropriations from the public Treasury in support of the National Museum, and of certain branches of work directly ordered by the Government itself. The library soon outgrew its limited quarters, and in 1866 was deposited in the Library of Congress, at a great saving of expense. The meteorological service was likewise transferred in 1874 to the Signal Corps of the United States Army.

For many years the institution conducted explorations in regard to the ethnology of the Indians of North America, and this has developed into an important Bureau of Ethnology, supported by Government appropriations, yet controlled by the Smithsonian.

The botanical collection was transferred to the Department of Agriculture, and the osteological specimens were placed in the Army Medical Museum.

The Smithsonian has been exceedingly fortunate in its executive officers. After the death of Henry, in 1878, Prof. Spencer F. Baird, the eminent naturalist, was called to the secretaryship. He had been United States Commissioner of Fishes for seven years and Assistant Secretary of the Smithsonian for twenty-eight years, and thus brought to the post wide experience as well as administrative ability. Under his care the National Museum was especially augmented, and the publications were issued uniformly on the lines laid down by his predecessor. Of his distinguished services to science we can not here take note; we merely quote two paragraphs from the resolutions adopted by the Board of Regents, November 18, 1887, on the occasion of his death :

" *Resolved*, That the cultivators of science both in this country and abroad have to deplore the loss of a veteran and distinguished naturalist, who was from early years a sedulous and successful investigator, whose native gifts and whose experience in systematic biologic work served in no small degree to adapt him to the administrative duties which filled the later years of his life, but whose knowledge and whose interest in science widened and deepened as the opportunities for investigation lessened, and who accordingly used his best endeavors to promote the researches of his fellow-naturalists in every part of the world.

" *Resolved*, That his kindly disposition, equable temper, single-ness of aim, and unsullied purity of motive, along with his facile mastery of affairs, greatly endeared him to his subordinates, se-cured to him the confidence and trust of those whose influence he sought for the advancement of the interests he had at heart, and won the high regard and warm affection of those who, like the members of this board, were officially and intimately associated with him."

Prof. Baird was succeeded in the office of Secretary by the present incumbent, Prof. Samuel P. Langley, LL. D., known to the scientific world by his masterly researches in solar physics. Under his administration the Smithsonian continues its pros-perity with undiminished vigor.

In a second article we shall consider the present status and many activities of this noble institution.

PART II.—ACTIVITIES OF THE SMITHSONIAN INSTITUTION.

IN our first article we attempted to show the circumstances which led to the founding of the Smithsonian Institution, to trace its growth, and to sketch the peculiar field which it occu-pies. The latter, however, can well be supplemented by a suc-cinct statement of its condition at the present time, or rather in 1895, the date of the most recent Annual Report.

MEMBERS OF THE INSTITUTION.—Presiding officer (*ex officio*), the President of the United States; Chancellor, the Chief Justice of the Supreme Court of the United States; the Vice-President of the United States; the Secretary of State; the Secretary of the Treasury; the Secretary of War; the Secretary of the Navy; the Postmaster General; the Attorney General; the Secretary of the Interior; the Secretary of Agriculture; the Secretary of the insti-tution.

ADMINISTRATION.—The business of the institution is managed by a Board of Regents, composed of the Vice-President and the Chief Justice of the United States, three senators, three members of the House of Representatives, and six other eminent persons nominated by a joint resolution of Congress. The Secretary of the institution is also secretary of this board and the principal executive officer.

BUILDINGS.—The Smithsonian Institution is housed in two buildings—the Norman, castle-like structure completed in 1855, and the huge one-story museum, to be noted below. The former is occupied as follows: The east wing contains the administration offices, comprising the rooms for the regents, the Secretary, the editor, and other officers. A small library of reference books (thirty thousand volumes) occupies a part of the ground floor. The main central hall is filled with valuable collections in ornithology and conchology, including the Isaac Lea cabinet of shells. Above this, another large hall is devoted to prehistoric anthropology. The west wing contains ichthyological specimens, and a very beautiful collection of crustacea, batrachia, and ophidia. In the south porch is a small group of instruments of research.

CORRESPONDENCE.—The official and casual correspondence of the Smithsonian Institution is no insignificant part of its daily life. Letters are addressed to the Secretary by the most learned scholars of Europe as well as by the humblest seeker after truth living in the wilds of North America, and all receive consideration and respectful answers. Tens of thousands of letters are annually received and acknowledged. If inquiries are made which the Secretary and his aids can not immediately answer, the letters are referred to eminent specialists outside of the institution.

The official list of correspondents, embracing learned societies and men of science throughout the world, numbers twenty-four thousand (1894). For a great many years the responsibility of the official correspondence devolved on the chief clerk, Mr. William J. Rhees, who is now keeper of the archives of the institution.

THE INTERNATIONAL EXCHANGE SERVICE.—The Board of Regents in 1851 established a system of international exchanges of the transactions of learned societies and of certain other classes of scientific works. The exchange extends also to specimens in natural history. In 1867 Congress imposed upon the institution the duty of exchanging official documents printed by order of either House, or by the United States Government bureaus, for similar works published by foreign governments.

This international exchange is of the greatest service to learned societies on both sides of the ocean, and to individual men of science who avail themselves of its privileges; it involves a prodigious amount of well-directed labor, as shown by the fact that in the twelve months 1892 to 1893 over one hundred tons of books were handled; these comprised 29,500 packages and 31,850 Government documents *sent out*, besides 101,000 packages and 5,196 Government publications *received*.

PUBLICATIONS.—There are three distinct sets of publications issued as serials, directed by the Smithsonian Institution:

1. Smithsonian Contributions to Knowledge, a quarto series begun in 1848, and comprising thirty-two volumes to date. In

United States National Museum.

these volumes are placed the monographs, articles, and papers offering positive additions to human knowledge, either undertaken by agents of the Smithsonian Institution or by persons encouraged by its assistance. These contributions correspond to the more elaborate memoirs of learned societies, and comprise treatises on anthropology, astronomy, biology, chemistry, electricity, ethnology, geology, mathematics, meteorology, natural history, palæontology, physics, and zoölogy, in all their ramifications.

2. Smithsonian Miscellaneous Collections, begun in 1862, thirty-five volumes, octavo. These contain bibliographies, tables, proceedings of Washington societies, and papers on scientific topics of value to scholars, yet not forming, as a rule, positive additions to the sum of human knowledge. These papers vary in size from a leaflet of four pages to a stout volume of twelve hundred pages. The individual articles are first issued independently, each receiving a number in course, and afterward they are bound up in volumes of suitable size, which themselves also bear numbers. This plan of publication also applies to the Contributions. The editorial work of this and the preceding series was long under the care of the late Mr. William B. Taylor, whose great erudition and skill in book-making proved invaluable to the institution.

3. Annual Reports of the Board of Regents—forty-nine volumes, octavo. These are submitted to Congress, in accordance with a clause in the act of incorporation. They contain the Journal of the Proceedings of the Board of Regents, the Report of the Executive Committee, the reports of the Secretary and of the directors, curators, or managers of the important departments controlled by the institution. In these reports are exhibited the financial affairs of the institution, its condition, its operations, and statistics of every kind connected with the same. Following the official part is a General Appendix containing a selection of memoirs of interest to collaborators and correspondents of the institution, teachers, and others engaged in the promotion of knowledge. These essays are generally reprints from divers sources, but they also include original translations and occasionally contributed articles. From 1880 to 1889 this General Appendix was chiefly devoted to an Annual Record of Scientific Progress prepared by specialists.

LIBRARY.—By exchanging the publications of the institution for transactions of learned societies, and for productions of foreign scholars, as well as by purchase, a library has been gathered of enormous value, now numbering over three hundred thousand titles. As already stated, it is merged in the Library of Congress, with the exception of a small collection for the use of the officers, partly housed in the Norman building and partly in the Museum. In the magnificent library building now approaching completion on Capitol Hill, the Smithsonian will have a separate hall for its deposit.

THE NATIONAL MUSEUM at first occupied the larger halls in the Norman building, and since 1858 special appropriations have been made by Congress for its maintenance; but, outgrowing its quarters, an independent building was erected by Congressional aid in 1881. This building has an available floor space of one hundred thousand square feet, but has been greatly overcrowded for many years. The director of the museum, who is also Assistant Secretary of the Smithsonian Institution, G. Brown Goode, LL. D., is assisted by thirty-three curators in charge of as many departments. These are: arts and industries, embracing twelve sections; materia medica, animal products, naval architecture, fisheries, foods, historical collections, coins and medals, transportation and engineering, Oriental antiquities, graphic arts, forestry, physical apparatus, helminthology, ethnology, American prehistoric pottery, prehistoric anthropology, mammals, birds, birds' eggs, reptiles and batrachians, fishes, vertebrate fossils, mollusks, insects, marine invertebrates, comparative anatomy, invertebrate fossils—paleozoic, mesozoic, and cenozoic—fossil plants, botany, minerals, geology. This mere catalogue of departments shows the prodigious range of subjects, the total number of specimens being more than three and a half millions. Nearly a quarter million of specimens were added in the twelve months ending 1892. The growth of the museum is due to many sources; these comprise the results of exchanges both abroad and at home, explorations by different departments of Government and by the Smithsonian Institution, collections secured through gift of foreign governments, and, most important of all, the collections obtained from several local and international exhibitions, in which the museum has always taken an active part.

An important activity of the museum is its generous distribution of duplicate specimens in natural history to scientific societies, colleges, and other educational institutions throughout the United States. Between 1871 and 1890, two hundred and seventy-eight thousand specimens were so distributed.

The museum is a favorite place of resort on the part of residents in, and visitors to, the Capital. In the year ending June 30, 1893, over three hundred thousand persons availed themselves of its privileges. Their examination of the objects is much hampered by the overcrowded state of the building, but it is assisted by the invariable courtesy of those in charge of the sections, and by books of educational value placed in the several departments. It rests with Congress to make an appropriation

INTERIOR OF THE UNITED STATES NATIONAL MUSEUM.

for erecting another building twice the size of the existing one, and only then will it be possible to display the treasures now stored in dark corners or still resting in unpacked cases.

The publications of the National Museum comprise two series: Proceedings of the National Museum, consisting of short essays giving accounts of recent accessions or newly ascertained facts in natural history, and promptly issued to secure the earliest diffusion of the information. These proceedings were begun in 1878, and are now comprised in seventeen volumes, octavo.

Bulletins of the National Museum, consisting of more elaborate memoirs relative to the collections, such as biological monographs, taxonomic lists, etc., varying in size from a few pages to many hundred pages. The bulletins were begun in 1875 and comprise fifty numbers to date.

THE BUREAU OF AMERICAN ETHNOLOGY was established in 1879, to conduct ethnological researches among North American Indians, and is supported by annual appropriations of Congress. The work is under the immediate direction of Major J. W. Powell, who was also a long time at the head of the United States Geological Survey, assisted by eminent specialists. The bureau conducts mound explorations, studies in ethnology, archæology, pictography, and linguistics of North America. Through its medium a wealth of information concerning the aborigines of North America is being treasured and made available to present scholars and to posterity.

The Bureau publishes four series of works:

1. Annual Reports, begun in 1879, now comprise twelve volumes, royal quarto. This series is handsomely printed and illustrated, and is both creditable to the Government and well adapted to attract public attention.

2. Contributions to North American Ethnology, begun in 1877; nine volumes, quarto.

3. Introductions to the study of various topics; begun in 1877; four volumes, quarto.

4. Bulletins; begun in 1877; twenty-six volumes, quarto.

THE NATIONAL ZOÖLOGICAL PARK.—From a desire to preserve certain American wild animals rapidly becoming extinct, living animals were exhibited in temporary quarters near the National Museum for several years. In 1889 the preliminary steps for the establishment of a Zoölogical Park were taken by the appropriation by Congress of two hundred thousand dollars for the purchase of land, and the park was actually founded by an act dated April 30, 1890, providing for the "organization, improvement, and maintenance" of a National Zoölogical Park. This act places the park under the direction of the Smithsonian Institution, and orders that it be administered for the advancement of science and the instruction and recreation of the people.

As soon as surveys could be completed, about one hundred and seventy acres of ground most picturesquely situated on Rock Creek, near Washington City, were secured and preparations begun for the reception of animals. This undertaking is so recent that little more has been accomplished than constructing roads, building animal houses, fences, etc., but there are already more than five hundred animals in the embryo Zoo. The natural features of the region, with its watercourses, ravines, rocky cliffs, forest trees, open glades, and sunny southern slopes, are superior to any site occupied in this way abroad or at home, and its extent is ten to fifty times greater than that of most of the gardens of Europe. Under the management of Dr. Frank Baker, the future of the National Zoölogical Park is very great; he plans to place the animals on ground appropriate to their natural habits and instincts, so that they can live under conditions similar to those enjoyed in freedom—a scheme only possible in a park of such great extent and variety of natural features.

G. BROWN GOODE.

ASTRO-PHYSICAL OBSERVATORY.—Prof. Baird had begun preparations for the establishment of an observatory for the study of the physical condition of celestial bodies, and when Mr. Langley succeeded to the secretaryship this eminent authority on solar physics soon secured its endowment by Congress. The late Dr. J. H. Kidder bequeathed five thousand dollars for prosecuting physical researches, and Dr. Alexander Graham Bell presented the like sum to the Secretary for the same purpose. In 1889-'90 a temporary wooden building was erected in the Mall south of the Norman building, and, though not entirely suitable for delicate research, much excellent work has been accomplished. In it are placed a Grubb siderostat, a spectro-bolometer constructed by Grunow & Son, and a galvanometer. These instruments, in the hands of Prof. Langley, are producing remarkable results, considering the inferior building and unsatisfactory site. It is to be hoped that these conditions will speedily be improved through Congressional appropriations.

HODGKINS FUND AND PRIZES.—Previous to the year 1891 the Smithsonian Fund had received only two small additions by gifts or bequests: one thousand dollars from Mr. James Hamilton in 1875, and five hundred dollars from Mr. Simeon Habel in 1880. In the year 1891, however, Mr. Thomas G. Hodgkins, of Setauket, N. Y., made the handsome donation of two hundred thousand dollars to the general fund, with certain conditions. In the formal statement of Mr. Hodgkins, dated September 22, 1891, he used these words: "This fund, to be called the Hodgkins Fund, and all premiums, prizes, grants, or publications made at its cost, are to be designated by this name; the interest of one hundred thousand dollars of this fund to be permanently devoted to the increase and diffusion of more exact knowledge in regard to the nature and properties of atmospheric air, in connection with the welfare of man in his daily life, and in his relations to his Creator, the same to be effected by the offering of prizes, for which competition shall be open to the world, for essays in which important truths regarding the phenomena on which life, health, and human happiness depend shall be embodied, or by such other means as in years to come may appear to the Regents of the Smithsonian Institution calculated to produce the most beneficent results."

To carry out the wishes of the donor, the following provisions for prizes, essays, and the Hodgkins medal were adopted by the institution, and announced in a circular issued in March, 1892:

J. W. POWELL.

1. A prize of ten thousand dollars for a treatise embodying some new and important discovery in regard to the nature or properties of atmospheric air.

2. A prize of two thousand dollars for the most satisfactory essay upon (a) the known properties of atmospheric air, considered in their relationships to research in every department

natural science, and the importance of a study of the atmosphere, considered in view of these relationships; (*b*) the proper direction of future research, in connection with the imperfections of our knowledge of atmospheric air, and of the connections of that knowledge with other sciences.

3. A prize of one thousand dollars for the best popular treatise upon atmospheric air, its properties and relationships (including those to hygiene, physical and mental). This essay need not exceed twenty thousand words in length; it should be written in simple language, and be suitable for publication for popular instruction.

4. The Hodgkins medal of the Smithsonian Institution will be awarded annually, or biennially, for important contributions to our knowledge of the nature and properties of atmospheric air, or for practical applications of our existing knowledge of them to the welfare of mankind. The medal will be of gold, with a duplicate in silver or bronze.

The treatises may be written in English, French, German, or Italian, and should be sent to the Secretary of the Smithsonian Institution before July 1,

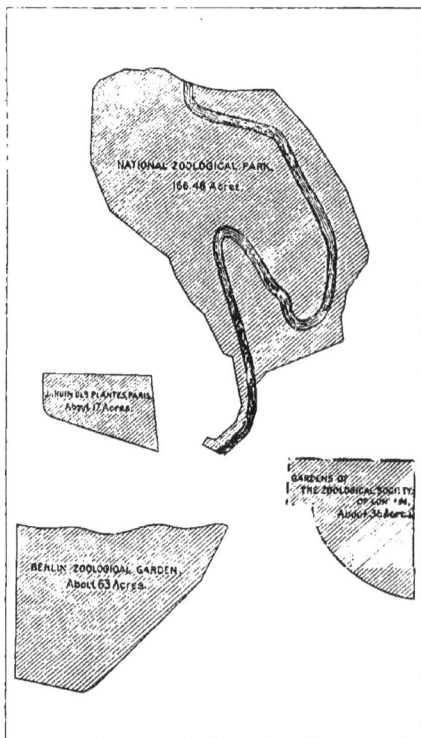

COMPARATIVE AREAS OF ZOÖLOGICAL PARKS.

1894; except those in competition for the first prize, which may be delayed until December 31, 1894. The time was subsequently extended to December 31, 1894, for all prizes.

Provision was made in the circular for a committee of award, for extending the dates above named, and for modifying the conditions prescribed. The circular also stated that special grants of money will probably be made to specialists engaged in original investigation upon atmospheric air and its properties.

In a supplementary circular, issued in April, 1893, it was stated that any branch of natural science may furnish subjects of discussion for the Hodgkins prizes, provided the subjects are related

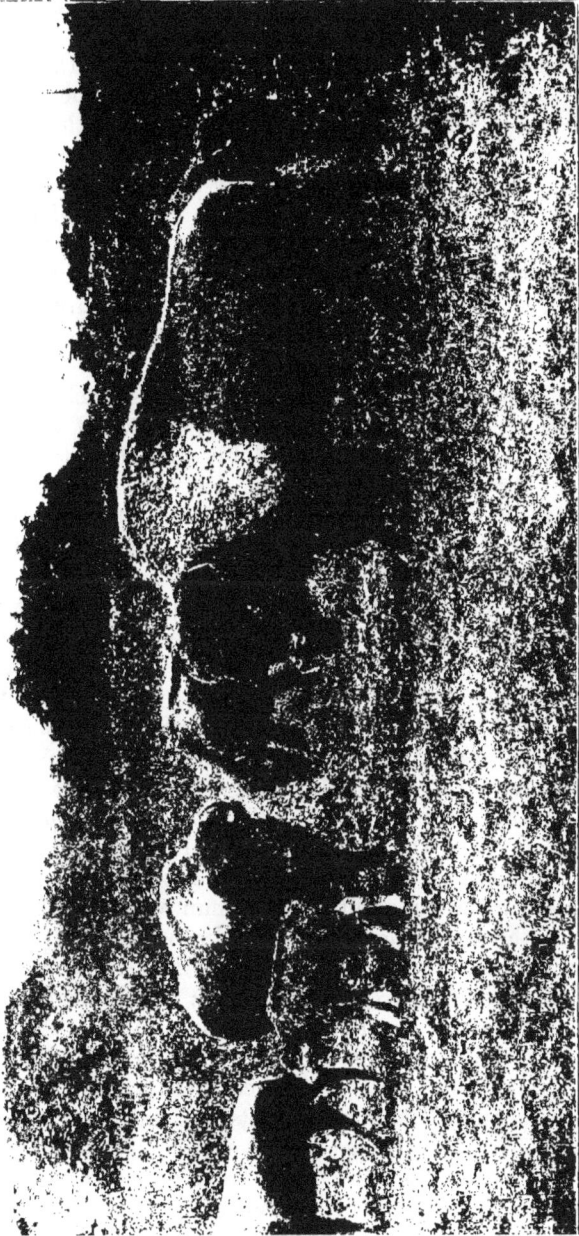

VIEW IN NATIONAL ZOÖLOGICAL PARK.

to the study of the atmosphere in connection with the welfare of men : " Thus, the anthropologist may consider the history of man as affected by climate through the atmosphere ; the geologist may study in this special connection the crust of the earth, whose constituents and whose form are largely modified by atmospheric influences ; the botanist, the atmospheric relations of the life of the plant ; the electrician, atmospheric electricity ; the mathematician and physicist, problems of aërodynamics in their utilitarian application ; and so on through the circle of the natural sciences, both biological and physical, of which there is perhaps not one which is necessarily excluded.

" In explanation of the donor's wishes, which the institution desires scrupulously to observe, it may be added that Mr. Hodgkins illustrated the catholicity of his plan by citing the experiments of Franklin in atmospheric electricity and the work of the late Paul Bert upon the relations of the atmosphere to life as subjects of research which, in his own view, might be properly considered in this relationship."

Eight thousand copies of these circulars were sent to institutions and investigators throughout the world, and applications for grants soon reached the Secretary of the Smithsonian.

In 1893 two grants were made : one of five hundred dollars to Dr. O. Lummer and Dr. E. Pringsheim, of the Physical Institute, Berlin University, for researches on the determination of an exact measure of the cooling of gases while expanding ; and a second grant of one thousand dollars to Dr. J. S. Billings, United States Army, and Dr. Weir Mitchell, of Philadelphia, for investigations into the nature of the peculiar substances of organic origin contained in the air expired by human beings, with specific reference to the ventilation of inhabited rooms.

Mr. Thomas George Hodgkins died November 25, 1892, at the advanced age of nearly ninety years ; being, next to Smithson, the most generous benefactor of the institution. A brief sketch of his life is appropriate. He was born in England in 1803, of highly respectable ancestry ; his early education was in France, where he acquired language, habits, and manners influencing all his later life. At the age of seventeen, led by a youth's love of adventure, and seeking relief from domestic unhappiness, he shipped before the mast on a trading vessel bound for Calcutta. The vessel was wrecked near the mouth of the Hoogly and the young man found himself penniless, friendless, and ill in a hospital in Calcutta. While in this sad plight, he made up his mind, so he said, to acquire a fortune and to devote it to philanthropic ends. After recovering he returned to England, then visited Spain, and after marrying in England he came to the United States in 1830. He immediately engaged in business and after

Deer House in National Zoölogical Park.

thirty years of successful ventures he retired on a handsome fortune. The fifteen years following this he spent in traveling over Europe and America, and in 1875 settled on " Brambletye Farm " at Setauket, Long Island, where he led a quiet, retired life.

For more than thirty years Mr. Hodgkins made a special study of the atmosphere in its relation to the well-being of humanity. He believed that this study was important, not only with reference to man's physical health, but even in relation to his moral and spiritual nature, and he hoped that the concentration of thought upon the atmosphere and its study from every point of view would in time lead to results which would justify his almost devout interest in the subject.

Mr. Hodgkins had no family and no known blood relations, and, recognizing the difficulties which often arise over the settlement of large estates, he chose to be his own executor; he therefore gave away his entire wealth to various public institutions; these gifts included large sums to the American Society for the Prevention of Cruelty to Children, and the similarly named society for protecting animals; and one hundred thousand dollars to the Royal Institution of Great Britain.

THOMAS GEORGE HODGKINS.

Since writing the foregoing pages, the Committee of Award for the Hodgkins Prizes has completed its examination of the papers submitted in competition. These papers were two hundred and eighteen in number, and were sent from almost every quarter of the globe. The committee consisted of Prof. S. P. Langley, *ex officio*, Prof. G. Brown Goode, Dr. John S. Billings, Prof. M. W. Harrington, together with a foreign Advisory Committee, composed of the late Prof. T. H. Huxley, M. J. Jansen, and Prof. Wilhelm von Bezold.

On August 6, 1895, the committee announced the following awards :

First Prize, of ten thousand dollars, for a treatise embodying some new and important discoveries in regard to the nature and properties of atmospheric air, to Lord Rayleigh, of London, and

Prof. William Ramsay, of the University College, London, for the discovery of argon.

Second Prize, of two thousand dollars, not awarded, owing to the failure of any contestant to comply strictly with the terms of the offer.

Third Prize, of one thousand dollars, to Dr. Henry de Varigny, of Paris, for the best popular treatise upon atmospheric air, its properties and relationships. Dr. de Varigny's essay is entitled L'Air et la Vie.

Besides these capital prizes, three silver medals and six bronze medals, coupled with honorable mention, were awarded to gentlemen for essays of great merit. To name all those awarded honorable mention would occupy more space than at our command.

On November 7th, Lord Rayleigh and Prof. William Ramsay called at the United States embassy, London, and received from the Secretary a check for ten thousand dollars, communicated by the Smithsonian Institution. It was a fortunate circumstance that the Smithsonian had the opportunity of awarding the first prize for so momentous a discovery as that of argon.

FINANCES.—The Annual Report of the Executive Committee of the Board of Regents for the year ending June 30, 1895, gives the following as the financial status of the institution:

Total Funds in 1895.

Bequest of Smithson, 1846....	$515,169
Residuary legacy of Smithson, 1867.	26,211
Savings from income, 1867.	108,620
Bequest of James Hamilton, 1875.	1,000
Accumulated interest of the James Hamilton fund, 1895.	1,000
Bequest of Simeon Habel, 1880....	500
Sale of bonds, 1881.	51,500
Gift of Thomas G. Hodgkins, 1891.	200,000
Residuary legacy of T. G. Hodgkins, 1894.	8,000
	$912,000

Receipts in 1894–'95.

Interest on fund, one year.		$54,473
Appropriated by Congress.	International exchanges.	17,000
	Bureau of Ethnology.	40,000
	National Museum.	166,500
	Astro-physical Observatory.	9,000
	National Zoölogical Park.	50,000
		$336,973

In addition to the above funds the Smithsonian Institution will soon receive the proceeds of a bequest made by the late Robert Stanton Avery, of Washington City, who died in 1894. The property bequeathed is estimated to be worth about seventy-five thousand dollars, and the income is to be devoted to special investigations in magnetism and electricity.

Finally, the position of the Smithsonian Institution is that of a "ward of the Government, having property of its own for which that Government acts as trustee, leaving its administration wholly with regents." Its most important function is to promote original research, reflecting thus the sentiment which occurs in the writings of James Smithson: "Every man is a valuable member of society who by his observations, researches, and experiments procures knowledge for men." The advancement of utilitarian interests commonly finds capital, for it appeals to the avarice of man; but the advancement of knowledge in its highest and widest sense secures little encouragement from wealthy men, and it is exactly this phase which the institution makes its own. Its next function is to make known to the world knowledge thus secured, for the benefit of mankind, and this it seeks to accomplish through its publications and their wide distribution.

The influence of the institution in local education is well shown by the following circumstance: Some years ago I was standing on the porch of the Norman building as two stout African "ladies" passed by. One of these remarked, "Let us go in there," pointing to the entrance. "Oh, no," replied the lady addressed, "there is nothing in there but 'Prehistoric Anthropology,'" pronouncing the words glibly and accurately. I listened with amazement, and pondered.